WHY ONLY TWO FPSOs IN U.S. GULF OF MEXICO?

The Late Start and Twenty Year Saga

Peter Lovie, Industry Pioneer
Oilfield Energy Center
2018 Hall of Fame Inductee

Disclaimer:
The information and images provided have been attributed as best as practically possible. Sources for content in tables are identified where known. Best efforts have been used to give a historical idea of what was done and sometimes corrobrated with recollections of former colleagues.

ISBN-13: 978-1-945532-82-5
Library of Congress Control Number: 2018953493
Printed in the United States of America

Published, Edited, and Cover Design by:

Opportune Independent Publishing Co.
113 N. Live Oak Street
Houston, TX 77003
(832) 263-1700
www.opportunepublishing.com

Peter M Lovie PE, LLC
P.O. Box 19733 Houston TX 77224

Peter.Lovie@FPSOsinGoM.com

An email to discuss your ideas or point out any errors or typos would be welcomed.

If you enjoyed this book and felt others might like it, please consider writing a review on www.FPSOsinGoM.com, Amazon, or on the website you purchased your copy.

WHY ONLY TWO FPSOs IN U.S. GULF OF MEXICO?

The upper image is of the FPSO *BW Pioneer*, owned by BW Offshore. It is the first FPSO in the U.S. Gulf of Mexico (US GoM) shown here while offloading to a shuttle tanker at the Petrobras operated *Cascade/Chinook* field development. First oil was 25 February 2012.
(Courtesy: BW Offshore)

The lower image is the second FPSO in US GoM: the SBM owned *Turritella*, shown on arrival at the Shell operated *Stones* development in US GoM in early 2016. Ownership has since been taken over by Shell.
(Courtesy: Shell)

Both FPSOs are discussed in the subsequent text, and the pictures repeated with more information.

Contents

List of Figures

Peter Lovie

List of Tables

Peter Lovie

Acknowledgements

The twenty-year saga for FPSOs in the GoM came about through the actions of thousands in the industry. The five individuals pictured here deserve particular recognition for their leadership: Allen Verret (Chevron, OOC, now retired), George Rodenbusch (Shell, now retired), Dave Bozeman (Devon, now retired), Cesar Palagi (Petrobras America) and Curtis Lohr (Shell, now retired). Their feedback has been invaluable in the preparation of this text.

For the historical background on the shuttle tanker business in Tables C and D, I am indebted to a former colleague: Alex Tischendorf, managing director for Teekay do Brasil Servicos Maritimos Ltda.

Peter Lovie

Preface

Introducing FPSOs into the U.S. Gulf of Mexico took much longer and was more difficult than anyone had imagined.

2016 saw the arrival of the second FPSO (floating production storage and offloading unit) in the U.S. Gulf of Mexico. Oil and gas industry experts recognized this as a major event because of the technical and business feats it accomplished. Although the U.S. Gulf of Mexico (GoM) has been a prominent pioneer for offshore oil and gas, FPSOs are the most used floating production system elsewhere and yet, the U.S. Gulf of Mexico has been curiously slow in adopting FPSOs.

The first FPSO arrived in 2010, but events beyond the control of the operating oil company led to first oil being delayed to February 2012.

This saga traces the events which made FPSOs in the GoM possible, starting in 1996 when GoM operating oil companies had already been producing offshore for fifty years. It tracks the steps from securing regulatory approval in principle, through the unexpected events which impacted the oil industry. It also explores the changes in design criteria and the decision-making process leading to the state of the industry in 2016, with one FPSO in operation and a second about to start production. Both employ unique Jones Act shuttle tankers.

Peter Lovie recounts frank anecdotes from the middle of it all. His unusual succession of Houston-based careers with an FPSO contractor, a shuttle tanker contractor, an operating oil company, and (currently) an independent consulting company give him an insider's insight on these events and a compelling outlook for future FPSOs in the GoM.

How Did It All Start?

It was not until fifty years after World War II that the idea of FPSOs in the U.S. GoM started to gain traction. Although the U.S. had started offshore production (most dramatically in 1947 with Kerr McGee's first platform out of sight of land), there had been only one U.S. FPSO precedent during that fifty-year period: an operation where Exxon's *OS&T* was moored in 490 ft. of water at their *Hondo* development offshore Santa Barbara, California. From 1981-1994, the location employed a 50,000 DWT tanker for production, with export by tanker. By comparison, elsewhere in the world, offshore Spain Shell's *Castellon* FPSO started operation in 1977, and on the Mexican side of the Gulf of Mexico, there had been occasional experiments with FPSOs since 1989.

It was not until 1995-1996 that Texaco's *Fuji* prospect drew attention to the possible use of an FPSO in the GoM with an unusual requirement in 1,700 feet of water— deep water in these days— and at a relatively remote location from pipelines. Back then, the leading regulatory authorities were the Minerals Management Service (MMS), concerned with safe petroleum production, and the United States Coast Guard (USCG), concerned with issues of vessel safety and operation. These realms were particularly relevant to FPSOs, which float, store, and offload oil.

In late 2001, the MMS Record of Decision stated that: "In

1996, Outer Continental Shelf (OCS) operators, as well as builders and operators of FPSO vessels, began having serious discussions with the MMS about the possibility of using FPSO systems in the Gulf of Mexico."

Operators in the U.S. GoM took the initiative to open discussions with the MMS. These talks revealed that MMS would require an Environmental Impact Statement (EIS) before they would consider any application to use an FPSO for any specific field development in the U.S. GoM. An EIS had not been required for earlier facilities in deep water such as jackets, guyed towers, tension-leg platforms (TLPs), spars, and semisubmersible production platforms; however, apparently what made all the difference was the FPSO's novelty, its storage of oil on the production facility, and its transport to shore by tanker instead of by pipeline. The *Valdez* oil spill had occurred just a few years earlier in 1989, and there was serious public concern over the potential of future spills.

Further, MMS did not see why they should undertake the effort of adopting FPSOs at their cost.

Preparation of an EIS meant a serious delay of at least two years. To any operator with the philosophy "time is money," it posed an excruciating obstacle to developing newly discovered reserves. This move eliminated the option of FPSOs from practical consideration. Other operators besides Texaco also wanted to have FPSOs in their toolbox for immediate action in future developments without the EIS delay.

Discussion among operators led the Texaco led DeepStar

industry group to step forward on behalf of the industry to fund and support the preparation of an EIS. It ultimately took more than $3 million from the operators to complete this venture: $1 million of cash funding to MMS plus in excess of $2 million in expenditures incurred by operators— no government handouts here!

Indeed, it was a huge challenge to mount the EIS effort to introduce FPSO developments to the U.S. Gulf of Mexico, which was accustomed to other developmental solutions that had multi-party support from both the government and private industry. Two engineers working with Texaco who understood the challenge the oil industry faced echoed Machiavelli's words on trailblazing in the figure below:

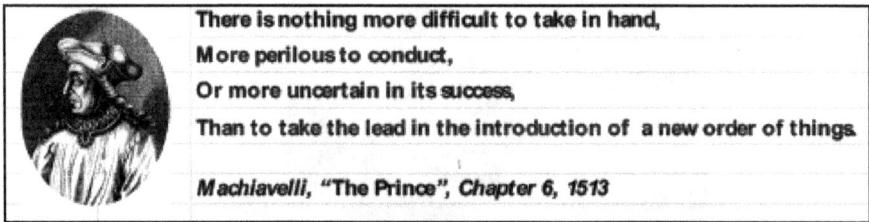

| There is nothing more difficult to take in hand, |
| More perilous to conduct, |
| Or more uncertain in its success, |
| Than to take the lead in the introduction of a new order of things. |
| |
| *Machiavelli, "The Prince", Chapter 6, 1513* |

<u>Figure 1</u>: Machiavelli's thoughts on the challenge of the EIS
(January 2000 SPE presentation, Jeffrey Harrison and Ron Skarbek of Texaco)

Under the leadership of Allen Verret, DeepStar campaigned directly through the Regulatory Committee of Deep Star and enlisted the support of interested operators, classification societies, and FPSO designers and contractors. These totaled about sixty organizations pulling together for the sake of field developmental options in deepwater GoM. Securing an EIS exceeded the initial projection of two years, taking close to five years.

During these years Allen Verret would joke about "cat herding" these various companies and interests. Although at times they pulled in different directions, they united in dedication to the same goal. In retrospect, it is difficult to imagine the EIS effort completion without Verret. He succeeded in balancing the technical, commercial, and political interests of more than 20 operators with those of regulators and FPSO contractors and designers, around 60 stakeholders in all.

Later Allen Verret served as executive director of the Offshore Operators' Committee (OOC) in the GoM, where he saw through the regulatory approvals for the first and second FPSOs in the GoM before his retirement in 2015.

Figure 2: Allen Verret, the Father of the EIS

In Metairie, Louisiana, countless meetings ensued at MMS offices. The subject of hot debate: requirements for FPSOs, which were established using information provided by the DeepStar team. Wanda Parker, Jeff Wolfe and the Offshore Technology Research Center (OTRC) were among many critical advisors. Part of getting the MMS to draft their Environmental Impact Statement was simply education on the potential of FPSOs.

The next challenge was identifying and filling gaps in regulatory processes and existing industry standards to cover an FPSO in the GoM. It was also necessary to assess the risk of possible spills, such as when oil would be stored offshore and transported to shore by tankers. Fortunately, the long history of lightering in the GoM and that experience transferring cargo between vessels of different sizes provided a guide on spill risks, though the rate of spills was very low indeed.

Because there were only three major designs of turret and swivel systems in use and each was highly proprietary, it was difficult to assess in detail the transfer of oil and gas from the seabed up into the FPSO. At the time of the EIS, there were still a few FPSOs around with drag chain turrets which had to be examined as well.

In public hearings at coastal locations along the Gulf of Mexico, members of the public could question what went on and had the opportunity to object to the directions being investigated in the EIS. As people educated themselves on what FPSOs were, how they might affect oil production, and how they differed from the hundreds of the familiar, existing production platforms, questions began to arise.

In one meeting in Lafayette, someone asked about the susceptibility of an FPSO to terrorist activity. This was obviously before 9/11, and the response was that the typical distance of an FPSO from shore, as well as the surveillance by industry and U.S. Coast Guard, would restrict movements of ships and small boats. It was not a clear-cut answer, but it was a fair judgment which drew from parallels with the North Sea, where the regulatory regime was also strict, and design criteria for storms were somewhat comparable.

At that time, as business development manager for North America on behalf of Bluewater of Holland I served as a spokesman for the use of FPSOs in U.S. waters. I clearly remember lugging around a six-foot-long model of a modern FPSO, which I heaved around between public hearings, from one city to the next, in a large, rented SUV.

The model was of *Glas Dowr*, a turret moored FPSO for the North Sea which Bluewater had designed, built, and operated. As an Aframax size of FPSO, *Glass Dowr* was relatively small and had the capacity to store about 600,000 bbl. Coincidentally, the first FPSO to start producing years later in the U.S. Gulf of Mexico was that size, even though the EIS suggested a larger Suezmax size of about 1,000,000 bbl.

Regulatory approval of FPSOs for the GoM necessitated that matters of petroleum-related production, vessel safety, and offloading be addressed. In a Memorandum of Understanding (MoU), USCG and the MMS negotiated and defined their scopes of work in how they would collaborate in approving different parts of the design, construction, and operation of FPSOs. On January 15, 1999, the MoU was published in the

Federal Register on pages 2260-2267 and was included in the EIS as Appendix A on pages 757-764. Fifteen years later, it was modified with the benefit of experience, remaining essentially the same in principle.

The approval effort for FPSOs also covered an approval process for shuttle tankers. The U.S. Coast Guard would largely regulate the delivery of crude oil production from FPSOs to refineries. There were two long-standing parallels of crude oil delivery by tanker from offshore:

(1) For more than two decades, the LOOP terminal offshore Louisiana (organized in 1972) had been importing more than a million bopd with more than 12 billion barrels to date, offloaded from the biggest of tankers: Suezmaxes, even VLCCs, and ULCCs. LOOP is 18 miles south of Grand Isle, La., in 110 ft. of water and is the only port in the U.S. capable of offloading deep-draft tankers known as ultra large crude carriers (ULCCs) and very large crude carriers (VLCCs).

(2) Even larger total volumes were brought to GoM onshore terminals from tankers too large to enter port via lightering operations using foreign flag tankers. These lightering tankers were about the size then contemplated for shuttle tanker service, and these operations had exemplary safety records.

Ultimately, Texaco's *Fuji* prospect was abandoned as subsequent appraisal wells indicated there was not enough oil there to make it economic. Nevertheless, this setback was hardly mentioned, as other operators remained as anxious as ever to proceed with the EIS.

It was the <u>operators</u> who chose the straw man configuration for the FPSO in the GoM: a storage capacity of <u>1,000,000 bbl</u> in a <u>double-hulled</u> vessel. Further, the operators stated that the FPSO would be <u>permanently moored</u> in the U.S. Gulf of Mexico.

Despite comments today, these key choices were not made by the regulators, although they certainly asked to have something realistic to work with. Later these criteria changed, but a start had to be made.

Shell's Investigation of an FPSO for the GoM Before Completion of the EIS

Shell conducted a remarkable investigation in 1998-1999 about the potential use of an FPSO as the development option for their *Na Kika* development in the U.S. GoM. It was a gutsy move before the EIS was complete, something I witnessed firsthand as a Bluewater employee while answering technical, operational, and project-related questions and proposing an FPSO as a solution.

Although the project eventually employed a semisubmersible instead, it was only after careful technical, operational, and economic assessments of an FPSO that Shell officials came to their choice.

George Rodenbusch at Shell led the deliberations on a project team numbering about fifty people, which included Shell and their non-operating partners in the development. Rodenbusch masterfully led and conducted this investigation of an FPSO independently from the EIS, a marvel of careful, point-by-point examination of design choices. One of these choices was the use of steel catenary risers (SCRs), for which we were not able to find a practical solution to justify an FPSO. Years later, Shell mastered the use of SCRs in deep water with an FPSO in Brazil in their BC-10 development.

Ultimately, in their 2016 installation at *Stones* in the U.S. GoM, Shell employed a version of the early lazy wave SCRs.

Still, this serious, objective, months-long endeavor put an end in my mind to the belief voiced by some that there was some kind of industry prejudice against FPSOs in the GoM. Although FPSOs were unfamiliar in the GoM and some operators repeatedly used spars (Anadarko) or TLPs (Conoco), their reasoning was for sound project factors: to mitigate risk, reduce cycle times, and accelerate returns.

Even Shell's development name was symbolic; *Na Kika* was an octopus that had arms stretching out in different directions, just like how subsea flow lines reached out to wells a distance from the central FPSO location. As with their careful consideration of options, the selected name was just one more sign of the thought that went into each choice.

Figure 3: George Rodenbusch of Shell: an Early FPSO Pioneer

Delivery of the EIS (2001)

The MMS published the Environmental Impact Statement in January 2001. Figure 4 below shows the front page of the 793-page document. The full EIS is available for download at http://www.bsee.gov/uploadedFiles/BSEE/Newsroom/ Publications_Library/2000-090.pdf

Before the EIS could be used in plans for FPSOs in the GoM, the U.S. government had to issue its formal Record of Decision, which it did on December 13, 2001, shown in Figure 5.

MMS' policy decision to accept the use of FPSOs in the GoM was subject to certain limitations, given on the 37th page of the Record of Decision.

The main limitations in the EIS were sensible:

+ The acceptance applied only to the central and western planning areas, where oil and gas leases were located;

+ The acceptance excluded designated lightering zones to avoid traffic hazards;

+ The acceptance applied only to water depths of more than 650 ft. because in shallow depths it was highly unlikely an FPSO would be competitive with fixed platforms;

+ As with oil delivered by shuttle tankers and gas by
pipeline, there was to be no flaring of gas.

In a press release on January 2, 2002, the MMS announced
it was ready to accept applications for the use of FPSOs in
the GoM. Moving forward, there was a new question: who
would go first? In September 2002 at SPE's FPSO Global
Workshop, operators, regulators, FPSO contractors, and
shuttle tanker contractors all voiced their views. A total of
154 paying attendees were in the conference hall, a figure
which commercial FPSO conferences in the Houston area
have not matched to this day.

Regarding what was now possible within the industry, most
attendees thought there might be four to six FPSOs in the
GoM within the next five to ten years. Little did we know
back then how slow the adoption of FPSOs would be!

OCS EIS/EA
MMS 2000-090

Proposed Use of Floating Production, Storage, and Offloading Systems On the Gulf of Mexico Outer Continental Shelf

Western and Central Planning Areas

Final Environmental Impact Statement

Author

Minerals Management Service
Gulf of Mexico OCS Region

Prepared under MMS Contract
1435-01-99-CT-30962

Cover

Turret-moored FPSO in a tandem offloading configuration with shuttle tanker (illustration courtesy of Advanced Production and Loading AS, 1999).

Published by

MMS U.S. Department of the Interior
Minerals Management Service
Gulf of Mexico OCS Region

New Orleans
January 2001

Figure 4: The Environmental Impact Statement (EIS)

_____ **Alternative B-4** (Approve the general concept of using FPSO's with a requirement for an attendant vessel.)

_____ **Alternative C** (No action at this time (insufficient information to make a decision)).

_____ **Other** _____

This decision, authorized by the signature below, and this Recommendation and Decision Document together establish the Agency's Record of Decision on the Environmental Impact Statement prepared on the Proposed Use of Floating Production, Storage, and Offloading Systems on the Gulf of Mexico Outer Continental Shelf, Western and Central Planning Areas. This programmatic decision is effective immediately. This decision does not constitute approval of any specific FPSO project. Submission, review, and approval of all required OCS plans, permit applications, and other submittals must be completed for every proposed FPSO system.

Dated: _13 December 2001_

Carolita U. Kallaur
Associate Director for
Offshore Minerals Management

Figure 5: Signature Page in Record of Decision Allowing
Consideration of FPSOs for GoM

In addition, SPE held another related event in 2003 on the topic of "Deep Water Oil Transportation." Discussion compared the existing pipeline network in the GoM with the potential use of shuttle tankers to bring production from FPSOs to GoM refineries. I organized and chaired both of these conferences but found interest in FPSOs steadily waned in the subsequent three years.

Despite the willingness of the MMS to accept FPSO applications, not one project came forward until years later in 2006. At industry conferences, I would run into Chris Oynes, the director of the MMS, and he'd ask: "Peter, when will we see the first FPSO application?". And each time I had to admit, I had no clue when it would happen.

While writing this, I sat down with Allen Verret (see earlier), and we concluded that we were all lucky the EIS happened when it did. In the current industry climate and post-*Macondo* regulatory climate and governmental management style, it would have been virtually impossible to accomplish the EIS approval in 2016!

Peter Lovie

Operators Consider FSOs in Addition to FPSOs (2000-2006)

In 2000, before the EIS release, a supermajor conducted a careful assessment on the use of floating storage and offloading units (FSOs) to store crude oil production from expected nearby spars, TLPs, and semisubmersibles. FSOs would enable flexibility of export to different GoM destinations, as well as better speed and economy. Also, FSOs would be an alternative to the construction of a traditional pipeline network in deep water. After months of vigorous, protracted debate which weighed engineering and economic factors, this supermajor decided to abandon the idea of the FSO. Nevertheless, many continued to see shuttling as an important export option.

Devon Energy had a similar idea in 2005-2006. The motives were similar: the ability to more economically and quickly develop nearby developments. In staying away from large long-term pipeline investments, they wanted to avoid the delays and costs of being beholden to pipeline companies. As a large and independent American company, Devon Energy had a portfolio that was, at one point, second only to Chevron's in the Lower Tertiary trend in deepwater GoM.

Figure 6: Dave Bozeman, Vice President at Devon Energy

After prolonged design studies, Devon patented its Floating Storage Offloading for the Gulf (FSOG) concept. Dave Bozeman (Figure 6) pioneered the studies on the use of FSOs, FPSOs, and shuttle tankers among multiple deepwater fields. He later collaborates with me in showing how shuttle tankers could be feasible and economical in the pipeline-dominated GoM in the ultra-deep, where so much of Devon's portfolio was located.

Ultimately, Devon's development plans came together in the deepwater prospect that became known as the *Cascade/ Chinook* development.

Shuttle Tanker Contractors Enter the U.S. GoM Market, Scratch for Business (2001-2005)

Use of shuttle tankers follows naturally from the use of vessels requiring offloading, i.e., FSOs and FPSOs. With the EIS in place at the end of 2001, the stage was set for developing shuttle tanker business to support expected FPSO installations. During this time, I started a new job in January 2002 as vice president of business development and corporate secretary with American Shuttle Tankers, LLC. (AST).

The remote and very deep waters of Alaminos Canyon had attracted industry attention. It was located offshore South Texas, close to the Mexican border, and contained blocks held by Chevron, Shell, Unocal, and others. It seemed an ideal location for an FPSO with export by shuttle tanker. Simultaneously, several other ultra-deepwater prospects in the U.S. GoM garnered operators' early attention.

These circumstances led two competing companies to propose shuttle tanker services for the U.S. GoM market: Seahorse Shuttling & Technology, LLC (which was a unit of Conoco) and American Shuttle Tankers, LLC (AST) that was owned 50:50 by lightering company Skaugen Petrotrans in Houston and shuttle tanker provider Navion of Stavanger, Norway. Both faced the same ground rules:

Peter Lovie

- Shuttle tankers in the GoM had to be very safe. In response to Exxon's1989 *Valdez* spill in Alaska, the United States Environmental Protection Agency had enacted the Oil Pollution Act of 1990 (OPA 90) legislation on tanker design and operation;

- In most cases, ports were restricted to a maximum draft of 40 ft., necessitating a design that was at best a small Aframax tanker or (more likely) a Handymax size of tanker;

- Tanker construction was far more difficult in the U.S. than anywhere else in the world because of fewer shipyards, slower delivery, and costs which were 2 ½ times more costly than the industry was used to seeing in the Far East!

- In the GoM, cycle times (pickup of cargo until delivery at a port) were similar to those in the North Sea, the birthplace of shuttling;

- Most importantly, the shuttle tankers had to be Jones Act-compliant: U.S.- built, U.S.-crewed and 75+% U.S.-owned.

The Jones Act heavily penalized shuttle tankers. Senator Wesley Livsey Jones (1863-1932), a Republican from the state of Washington, wrote the Merchant Marine Act of 1920 to protect his state's trade with Alaska. Decades later, although the protectionist days of the 1920s were long gone, the Jones Act continues to have powerful lobbyist support. Its effects were something that both shuttle tanker contenders had to face. They chose different solutions to the same problem.

Figure 7: Senator Wesley Jones of the 1920 Jones Act

The first was Conoco's subsidiary, Seahorse Shuttling & Technology LLC, which chose to create a new design tailored to GoM conditions with Korean ship construction input. To be built in a shipyard in Alabama, the ship would be able to enter GoM ports, operating with a draft of no more than 40 ft., dynamically position (DP2), with a bow loading system (BLS) and a storage capacity of 550,000 bbl of crude oil.

In the 1990s, Conoco had had extensive experience with shuttle tanker offloading in its field developments in the North Sea. They knew the history of the North Sea shuttle tanker business: in 1979, Statoil had pioneered the business. It was later hived off as a separate venture, Navion ASA, based in Stavanger. Teekay, a Vancouver-based major international shipping company, subsequently fully acquired Navion ASA in 2005. Conoco had brought together local GoM experience with the best available in shuttle tanker expertise from the harsher environment of the North Sea.

Figure 8: The GOMAX Shuttle Tanker Design From Seahorse Shuttling

The AST shuttle tankers would also be DP2, i.e., dynamic positioning but with a backup system for added reliability and minimal risk of spills— the best used in the shuttle tanker industry. The hull would be either an existing tanker design adapted to be a new build in a U.S. yard or a DP2 conversion of an existing product tanker of Handymax size.

The Jones Act requirement of 75+% U.S. ownership would be satisfied through a U.S. owned entity owning the shuttle tankers and bareboat chartering them back to AST on a long-term basis. Training of and management of U.S. crews would be to North Sea standards used by AST's Norwegian part owner (Navion) and by local owner (Skaugen).

AST started to examine Aframax solutions but found it more practical to adapt existing Handymax designs. They settled on existing product tankers with capacities around 330,000 bbl, which were to be converted with BLS and DP2 station-keeping. In outward appearance, they looked like the shuttle tankers ultimately contracted at *Cascade/Chinook* a few years later (Figure 13).

A common theme with both contenders was safety. They drew on their experience in the North Sea and attempted to emulate the same high operating standards which were based on technical criteria in order to operate with a high level of uptime in rough sea conditions.

During the late 1990s and early 2000s, management culture stressed more training and transparency had steadily lowered hazardous incidents. Systems that had proven to be safe in the harsh North Sea were to be adapted to the somewhat milder sea spectra of the U.S. GoM, an environment where everyone was likewise extremely sensitive to oil spill risks.

Whether in the North Sea or in the Gulf of Mexico, the precise control performance of DP2 tankers meant that the significant wave height (Hsig) limits for hookup and castoff could be reliably and safely set at relatively high levels to allow maximum up times.

A practical economic advantage of DP2 shuttle tankers was that they allowed the possible elimination of one (if not both) service vessels offshore at the FPSO's location, used as a hold-off tug and as a hose-handling vessel.

Another less-recognized value of DP2 in the GoM was the potential time saved and increased safety of maneuvering entry into ports and docking. In most instances, DP2 tankers could eliminate the need for tugs and pilots.

I witnessed these benefits firsthand in 2003 while riding one of Navion Shipping's DP2 Suezmax shuttle tankers from the *Big Stone* Anchorage in the Chesapeake Bay. We rode up the channel, past Fort McHenry that inspired the "Star-Spangled Banner," all the way up to a refinery in Philadelphia to deliver a cargo of North Sea crude. The captain was able to program the desired track in the channel within a foot or two. On arrival at the dock, the captain docked the ship by himself, operating controls standing on the wing. Crews of waiting tugs idled, wondering what was going on and whether they would be needed. I suppose the waiting terminal people must have wondered if the captain would screw it up and damage his ship and their dock, but all went off smoothly.

Many in the offshore community saw the advantages again in 2004 when another of Navion's DP2 shuttle tankers arrived from the North Sea in the Houston Ship Channel in order to deliver a cargo of condensate for Exxon at the Oiltanking terminal. The U.S. Coast Guard Captain of Port spoke glowingly of how this partially-loaded Suezmax tanker was maneuvered with great ease and safety. He relayed this fact to four busloads of industry visitors that came to see this vessel! At the time, some in the U.S. GoM felt that implementing North Sea practices was overkill for the GoM. Despite obvious benefits for safety and efficiency of operation, this belief persisted, focused mainly on whether or not to use DP2. Today, many adhere to the same wariness, despite the politics of extreme sensitivity to spill risks.

Now DP is widely used in cruise ships, which are often as large as Suezmax tankers. It is one of their secrets to quickly get in and out of ports and harbors in any part of the world.

During 2003 and 2004, AST saw the preponderance of spars and non-FPSO solutions in deep water. Therefore, AST attempted to adapt to the market by creating its Separate Storage Shuttling (S-S-S) system in which a DP2 tanker would be parked adjacent to a spar, semisubmersible or other non-storage facilities to store oil between offloadings to shuttle tankers. S-S-S was marketed widely to the GoM offshore community. A U.S. patent was filed, naming Peter Lovie as the inventor. At the 2004 Offshore Technology Conference (OTC) the concept was awarded a Meritorious Engineering Award (MEA) at the annual ceremony organized by Hart's E&P magazine. It made no difference—despite all the studies, creative thought, patent application, awards and, presentations, none led to any takers!

As 2004 progressed, the need for FPSOs and hence shuttle tankers became slim. Inquiries for shuttle tankers were few and far between, despite multiple deepwater developments on the horizon. Seahorse Shuttling succeeded in using shuttle tanker export as a competitive threat to negotiate down high pipeline tariffs for a field development operated by their parent company Conoco, which paid off many-fold according to street talk. They chose to accept this success, wind up their shuttle tanker company, and move on.

American Shuttle Tankers bid on one particularly serious project, Shell's *Great White* development. Earlier in the dealings, AST had talked about absorbing some project risk (as might be done in the North Sea market, which was able to pool tankers between multiple customers because of much larger fleets and broader demands for shuttle tanker services. However, for potential GoM shuttle tanker projects with small fleets, it became clear that it was not commercially sound for AST to make a contract of affreightment or take on residual risk. In other words, these North Sea business models would not work in the GoM, and the Shell commitment would have to be a full-payout deal, meaning significantly higher rates. It disappointed Shell: shuttle tankers would be less economically favorable. By that time, Teekay had acquired an interest in AST via Navion's 50% ownership of AST. Teekay decided it was not prudent to take the project risk associated with providing shuttle tankers for use with FPSOs in this new market of the GoM.

As events progressed, the *Great White* field developmental solution went in another direction. Instead of an FPSO, it became a spar, in what is now in production at *Perdido*. In 2005, AST also disappeared, absorbed into what is now Teekay Offshore Operators (NYSE: TOO). With operations in Brazil, Eastern Canada, and the North Sea, Teekay is one of the largest shuttle tanker owning and operating companies.

After AST's disappearance, there was no shuttle tanker contractor in the business of proposing services for the GoM.

As an aside, I could understand why Shell called the development Great White. In a video of one subsea well completion for the development, the team had observed a great white shark at an unusually great depth of several thousand feet. The name channeled the pioneering spirit needed to tackle that remote and very deep difficult development. However, when Shell later called the development *Perdido*, it ultimately did not live up to the imaginative naming previously seen with *Na Kika*!

Peter Lovie

The 2005 Hurricanes—Mother Nature Changes the Game

Just as the shuttle tanker contractors left the U.S. GoM, along came Hurricanes *Katrina* and *Rita* to screw things up for the offshore industry!

They were two of the worst offshore hurricanes in living memory, both in the same year, weeks apart. They damaged and destroyed numerous platforms and washed away seabed pipelines for miles. It took many months to repair the pipelines, interrupting delivery of oil and gas from non-damaged platforms, which could no longer produce.

MMS counted 113 production platforms destroyed and 52 heavily damaged. A number of jackup mobile offshore drilling units (MODUs) were destroyed, and around 19 jackup and semisubmersible MODUs went adrift. The devastation prompted the oil and gas industry to rethink design criteria for offshore installations.

The industry started to realize that the evidence was there to call for changes in metocean design criteria. Hitherto, we had simply underestimated meteorological and oceanographic criteria. At the Offshore Technology Conferences in 2006 and again in 2007, industry gurus had a field day in technical sessions as they codified criteria for future use.,

The choice of metocean design criteria affected the design of any FPSO for the U.S. GoM and how it might be moored. It was now an urgent matter; MMS was then considering the expected FPSO at the *Cascade/Chinook* development for Deep Water Operational Plan (DWOP) application. Petrobras America had recently taken over operatorship of this development from BHP. *Cascade* was now 50-50 Petrobras and Devon, and *Chinook* was 67-33 Petrobras to Total.

Figure 9; Typical Pipeline and Platform Damage From Hurricanes *Katrina* and *Rita* in 2005.

Bringing Production Back on Line—The First DP Shuttle Tanker in the U.S. GoM

After all the damage, it became a national emergency to bring offshore oil and gas production back online. With its marine expertise and previous experience in the North Sea, BP decided to bring in a DP shuttle tanker to get one of their offshore locations going again. At this location, the collection of condensate was particularly important to enable resumption of full gas production.

Part of the challenge was getting a North Sea shuttle tanker over to the GoM on short notice and lining up the necessary hoses and dry cutoff equipment.

The bigger challenge was political; because of the Jones Act, it was no easy task to negotiate the quick transport of liquid cargo to a U.S. port. Arranging a Jones Act waiver meant getting it signed by the U.S. president.

BP's *Nordic Trym*, shown in Figure 10, became the first and only DP shuttle tanker to operate in the U.S. GoM!

Figure 10: Teekay's *Nordic Trym*: The Only DP Shuttle Tanker to Ever Operate in the GoM

The "Aha Moment" for FPSO Disconnection in the GoM

The hurricanes had caused a record number of MODUs not just to drag anchors but be set completely adrift. Although this had happened before, it had never happened on this scale.

A single-column TLP at the *Typhoon* development was capsized (some believed by a collision with one of the drifting MODUs, although that was never conclusively proven).

It started to dawn on operators that there was a collision hazard which had not been anticipated in their planning and design. Not only that, certain operators were already talking of installing FPSOs that might have half a million barrels of oil stored in their hull, possibly more.

A hypothetical specter arose: a 40,000-to-50,000-ton drifting semisubmersible could slam into a crude oil-laden FPSO during or after a hurricane and break open the FPSO's hull, causing a major oil spill. The potential of a horrendous risk and liability emerged, maybe bigger than a *Valdez*! No one back then could ever conceive what might happen in *Macondo*, but the situation was a potential nightmare however you looked at it.

The number of MODUs that broke loose outnumbered the total that experienced total mooring failures and drifted a significant number of miles. However, the number of total mooring failures was still serious, as Table A shows.

MODUs adrift in 2005 hurricanes: Total mooring failure		
2005 Hurricane	Total MODUs	Distances drifted in the GoM
Katrina	5	between 4 and 80 miles
Rita	7	between 75 and 145 miles
Source: MMS/OTRC		

The total MODUs adrift with partial mooring failure was much larger, but the distances they moved was much less.

Table A: The inspiration Behind FPSOs Being Disconnectable in the GoM!

Many of these MODUs had been working in deep water, right where a future FPSO might be expected. This was no unique coincidence, and it called for some kind of countermeasure.

At the start of 2006, GoM operators had a collective epiphany. Perhaps "epiphany" is not the right word. It was as if all the GoM operators in Houston said in unison, "Holy crap, we have to make our FPSOs disconnectable."

It was pretty sensible. No one could ever take that oil spill risk. Despite the small probability, the risk had horrendously large consequences. It was not the regulators who came up with the requirement for FPSOs to be disconnectable. Once again, the operators took the lead.

Despite all the bad press often attributed to Big Oil, this event was an example of oil companies acting responsibly—and doing so without any government decree!

Today, people are starting to forget the events that unfolded. Some consultants and contractors question disconnectability and think they can do better without bothering with it. However the small probability of a massive spill due to a MODU collision still weighs with operating oil companies that make their income from producing oil. MODUs may be better moored nowadays, so the collision risks may indeed be less in future years. But savvy operators remember and weigh the risks in a post-*Macondo* world.

In the Far East, disconnectable FPSOs had served as an escape from severe cyclonic storms, typically offshore northern Australia and in the South China Sea. It had allowed design criteria for the FPSO to be less demanding. By 2006, many FPSOs (such as the FPSO *Glas Dowr*) had been designed to be permanently moored in the North Sea with extreme conditions not too different from the U.S. GoM hurricanes. Several had worked there during the prior decade. Therefore, survival in the U.S. GoM hurricanes would have indeed been doable with a permanent mooring system, just as the EIS had foreseen years earlier, had it not been for the collision risk.

It became obvious that the ultra-deep waters of the U.S. GoM necessitated careful forecasting and disciplined integration with disconnection preparation. With ultra-deepwater systems of risers and mooring lines, the requirements differed from earlier disconnectables in the Far East.

Disconnection posed one side benefit: to modify topsides if necessary more readily. Rather than having to modify offshore, as would be necessary with a permanently-moored FPSO, disconnectable FPSOs allowed modifications to be done dockside nearby.

Settling on a Contract for the First FPSO (2006-2007)

From 2006 to 2007, it was a difficult time to put a contract together for the first FPSO. In 2006 the oil industry was revising GoM design criteria that had been established years before. With the first contracting steps for the FPSO anticipated in 2007, these revisions were cutting it close. Simultaneously, contracts had to satisfy regulators and assure them that all would be well in the world of revised metocean criteria. All parties had to be responsibly compliant with these new requirements. Such were the challenges faced by Petrobras America as the operator on a pioneering venture to install an FPSO at their proposed *Cascade/Chinook* field development.

An FPSO had become an attractive option for the location in Walker Ridge block 249 in 8,200-ft. water depths to serve both the *Cascade* and the *Chinook* fields. Here the formations did not have good production analogues, and there was a risk that if formations did not produce as hoped for, the field development might not work out well. On the larger *Cascade*, the partners were 50-50 Devon and Petrobras with Petrobras as the operator. These partners aligned in their plans to test the development and taking action now. This stood in contrast with supermajors, which typically executed plans only after prolonged studies. Rather than a full field development solution, the FPSO at *Cascade/Chinook* was seen as an Early Production System (EPS) for a five-to-eight-year time scale. Later, according to production results, the FPSO could potentially be retained, replaced with something different, or even removed.

In 2006, DeepStar had discussed the use of a dynamically positioned FPSO in the GoM, but this idea came with a hazard: if the DP systems drove off, wells might be lost. A DP FPSO was not so bad for a single-well production scenario but was not advisable for the multiple wells anticipated at *Cascade/Chinook*.

The partnership between Petrobras, the national oil company of Brazil, and Devon, one of the large U.S. independents, struck quite a contrast in management styles. However, it was a great balance for the first FPSO in the GoM. Worldwide, Petrobras employed a larger fleet of FPSOs than Devon did and claimed much more FPSO experience. Devon had just hired Peter Lovie as senior advisor, floating systems. I reasoned that the lease rate of the FPSO nominated for *Cascade* was seriously over the market. There was a delay of several months as the discussion continued and a bid process opened up for the FPSO contract.

The five bidders for the FPSO lease contract were a different lot from what might have been expected a decade earlier: Bluewater, BW Offshore, Modec, SBM, and Teekay. In 1996, the Big Three FPSO contractors had been Bluewater, Modec, and SBM, Back then, Teekay was not in the FPSO business at all, and BW Offshore was a small player. Other new names joined that league table as the FPSO business expanded worldwide. By 2007, the Big Three had become BW Offshore, Modec, and SBM, a lineup that continues today.

The final competition for the first FPSO in the U.S. GoM ended up being a close race. BW Offshore finally drew ahead by a nose to win the contract to design, build, and lease the first FPSO in the Gulf of Mexico for a five-year term with three one-year options. It was a disconnectable FPSO designed by BW Offshore, as shown in the artist's impression in Figure 11.

Figure 11: Contract Signed With BW Offshore, August 2007 for First FPSO in the U.S. GoM

(Source: BW Offshore)

An unanticipated complication emerged as the FPSO bid process progressed. In most FPSO projects, the provision of export tankers means one of two things. Either the project implements a business model which employs well-vetted spot market "tankers of convenience," or it makes a longer-term commitment for shuttle tankers. In both options, export tankers are usually not on the critical path for a field development project. That changed in U.S. Gulf of Mexico with shuttle tanker deliveries for the project at *Cascade/ Chinook*. Thus, the overall timing of production from the entire field development became dependent on the shuttle tankers being able to start when promised.

With the Jones Act tankers in the U.S. GoM for *Cascade/ Chinook*, export tanker delivery times were on the critical path. It was just another surprise that Cesar Palagi (shown in Figure 12) and his colleagues at Petrobras America had to deal with. Palagi was responsible for his company's design and its implementation of development projects in the ultra-deep waters in Lower Tertiary fields in the GoM. He worked in a time of great change as a Brazilian team of experts joined the expanding Houston office.

The final design on the shuttle tanker was based on something Petrobras had pioneered in Brazil in the 2000s. Although it was not dynamic positioning, it did have enhanced maneuverability from its bow thruster and a variable pitch propeller, as well as a catenary hose arrangement with a bow loading system (BLS) of the type pioneered with the DP shuttle tankers of the North Sea. In essence, it was a compromise between the ultimate state of the art developed for the sensitive environment and regulatory regime of the North Sea - akin to what had been proposed four years earlier for the U.S. Gulf of Mexico - and what had been developed by Petrobras for the environment and regulatory regime of Brazil.

Figure 12: Cesar Palagi: Walker Ridge Production Asset
Manager with Petrobras America

In the 2000s, Navion had chartered North Sea shuttle tankers
for use in Brazil. Parallel local efforts created a concept to
offloading from FPSOs in Brazil. This became the basis of
what Petrobras chose for the U.S. GoM.

The choice of the offloading system at *Cascade/Chinook*
is thus quite unique. Made in the face of a demanding U.S.
GoM regulatory environment and the difficult consequences
of the Jones Act requirements, Petrobras also had to consider
the small size of tankers (by international standards) that are

Peter Lovie

needed to enter GoM ports. Industry scrutiny was intense, as it was for the first FPSO in the U.S. GoM.

To offer a safe operation, Petrobras America had steered a path between economy, relatively straightforward equipment availability and careful definition and management of procedures. This was accomplished in time charters for two shuttle tankers between Petrobras (NYSE: PBR) and Overseas Shipholding Group (NYSE: OSG).

The shuttle tanker configuration was thus different from proposed by either of the two GoM shuttle tanker pioneers of 2001-2005, and different from what the two non-operating partners at *Cascade/Chinook* (Devon and Total) were used to. It was a pragmatic choice in that the charter period was relatively short for this specialized type of vessel (5+1+1+1 years to match the terms of the FPSO lease), all this in the Jones Act world that often had longer-term commitments for newbuild tankers.

Petrobras' non-operating partner (Devon) had internally questioned the solution but had, on careful consideration, recognized that it came down to being a matter of safety. Dave Bozeman, my boss at Devon, would joke that "If they want to offload to barrels in a canoe, that's fine as long as they do it safely."

Commercial Basis—Time Charters for the First Shuttle Tankers in the U.S. GoM

Simultaneous with the FPSO bid competition, the shuttle tanker competition opened up. In other parts of the world, this bid process might not have been so difficult, but in the U.S. there was the Jones Act to contend with. The results: very few normal choices with CAPEX and OPEX numbers at 2.5+ times the world market that the development's partners were used to.

Petrobras America did a remarkable job in sifting through the various offerings, which included a variety of creative solutions from a variety of contractors, such as 30+ year-old tankers to be refurbished, articulated tug barges, and conversions of newbuild tankers that ultimately won the day, time chartered from Overseas Shipholding Group (OSG), one of which is shown in Figure 13.

Figure 13: August 2007- Petrobras Signs Time Charter With OSG for two Shuttle Tankers

A Parallel Investigation of Shuttle Tanker Export

Safety demanded that a ship-owning and managing organization be dedicated to the whole cycle through experience, training and company culture. Some felt the North Sea solution was safer than what is used at *Cascade/ Chinook* when one looks at connection/disconnection criteria in Table B. North Sea practice was much more restrictive than the Brazil practice: connection and disconnection wave height criteria were lower.

The other side to that argument is how the *Cascade/Chinook* system uses a larger separation distance between FPSO and shuttle tanker, plus a hold off tug with a slack hawser connection to the stern of the shuttle tanker, all as safety measures. So the overall safety comparison is complex.

The imponderable was the true relative risk levels of these different practices.

The significant wave height (Hsig) data for connect/disconnect show how the simple, well-tried floating hose offloading system could be practical in the U.S. GoM, although it risks a little more downtime than with the other two methods. The floating hose offloading system had already been used for years at the CALM (Catenary Anchor Leg Mooring) buoys at the Louisiana Offshore Oil Port (LOOP) system to offload

61

imported oil.

Origin of safety criteria	Shuttle tanker configuration for US GoM	Maximum wave height for _connection in US GoM_: Hsig, meters	Maximum wave height for _disconnection_ in US GoM: Hsig,
North Sea practice	Standard Handymax tanker, floating hose	1.5	2.5
	DP2 Handymax tanker, BLS	2.0	3.0
Brazil practice	Standard Handymax tanker, enhanced maneuverability + BLS	2.5	3.4

Definitions	
BLS	Bow loading system - uses catenary hose beteen FPSO and shuttle tanker
DP2	DP = Dynamically positioned with bow and stern thrusters and controllable pitch propeller, 2 = double redundancy
Enhanced maneuverability	Main changes from a standard tanker were to use a controllable pitch propeller plus bow thruster. Not DP. Developed in Brazil
Handymax	A tanker of around 330,000 bbl storage capacity. In the North Sea and Brazil, larger shuttle tankers are usually used, either Aframax (~650,000 bbl storage capacity) or Suezmax (~1,000,000 bbl)
Hsig	Signficant wave height. Hsig is easy to estimate at sea. One third of waves are higher, hence _maximum_ trough to crest wave height is higher than shown above.

<u>Table B</u>: U.S. GoM Shuttle Tanker Connection/Disconnection Criteria

For its own prospects, the design and operating philosophy at Devon had centered on finding the safest possible solution that was also thoroughly proven. We looked to the North Sea and how practices there could be adapted to the U.S. GoM. It was nothing original, exactly what Conoco and AST had promoted back in 2001-2005.

However, there was a serious practical difficulty: building DP2 tankers in the U.S. would be a special commitment. With no U.S. shipyards offering to build such vessels, the venture was unlike the experience of shuttle tanker owners in Europe, who could decide on the requirements they needed and quickly order what they wanted from shipbuilders in the Far East. Attempting to do a North Sea clone in the U.S. GoM for a limited life project of an Early Production System akin to *Cascade/Chinook* was commercially too difficult.

In the next two years, separate from their interest in the *Cascade* field development, Devon Energy (NYSE: DVN) went to some length to examine the potential for shuttle tankers for their prospective U.S. GoM developments. The results of their examination were shared in 2009 in a paper presented at the Deep Oil Technology conference (DOT 09). In general, these studies reported that in these remote locations, the use of an FSO alongside a facility without storage led to export economics about the same as those with a pipeline. In contrast with an FPSO used for a full field development, e.g., fifteen to twenty-five years field life, the export cost with DP2 shuttle tankers was about half that of pipelines.

As it ultimately played out, Devon exited the offshore business in 2H2009 to concentrate on onshore shale business, and there was no opportunity to work these studies on export from offshore fields through to a conclusion in reality! ("Export" here means the transport of crude oil production from the offshore fields to refineries, i.e., quite different from the very recent change in U.S. law to allow export of U.S. crude to other countries.)

Peter Lovie

Time to Modify or Repeal the Jones Act?

However one argues the DP v. non-DP case, the Jones Act has prevented any experienced shuttle tanker company from entering the GoM market and offering operational safety and efficiency at better terms with their foreign-built and foreign-flagged vessels.

Supporters of the Jones Act argue that it offers a reservoir of talent and vessels, which are valuable in times of threats to national security. There is a case to be made nowadays that it would be best to dispense with the part of the Jones Act which requires vessels to be built in the U.S. while keeping its mandate for flagging. The idea of U.S.-built vessels was much more practical fifty to seventy years ago when the U.S. was still a shipbuilding force to be reckoned with, or even further back in the era when Senator Jones was alive.

Unlike the U.S. marine transportation industry, the airline industry in the U.S. is free to employ airliners built by Embraer in Brazil, Bombardier in Canada, and Airbus in Europe, in addition to U.S.-built Boeing airliners.

For ground transportation, if you don't want to be "Ram Tough" or drive a Mack or GMC truck, you can choose to buy a Mercedes or Volvo or Isuzu truck!

If similar allowances were applied to marine transportation, getting vessels from U.S. port to U.S. port would become significantly easier. If the part of the Jones Act which mandated vessels be U.S.-made were repealed, it would radically change procurement of tankers for crude oil and products transportation and considerably improve economics. But right now that looks like a tough and time-consuming task of the kind that the engineers at Texaco had understood in 1996 in citing the wisdom of Machiavelli (see Figure 1)!

Similar arguments can be made for modifying or even repealing the Jones Act for other U.S. GoM transport-related activities, such as moving supplies from shore to MODUs or production platforms. Then there is the high cost of tanker delivery of diesel and gasoline from refineries in the U.S. GoM to the East Coast. It can cost half as much to move the same product from a U.S. GoM refinery to a more distant Canadian port with much more economical foreign flag tankers.

It all Hits the Fan (2010)

For a while, everything seemed fine. The Petrobras America team had successfully managed their way through the procurement and construction of the first FPSO for the U.S. GoM. The MMS and USCG had given their regulatory approvals for the operation of the FPSO, which would be located at a world record-breaking water depth. BW Offshore had completed the *BW Pioneer* FPSO in Singapore and sailed it under its own power to the Gulf of Mexico.

Within days of *BW Pioneer*'s arrival in late February of 2010, the *Macondo* disaster happened at BP's *Macondo* prospect. With around 4.9 million barrels of total discharge, this oil spill was an industrial disaster the likes of which we had never seen. Everything shut down.

The FPSO team at Petrobras America persevered. They found a way through all the obstacles the incident brought. Requests followed to use the freshly-arrived *BW Pioneer* in *Macondo* recovery efforts. Then the first shuttle tanker showed up and had to be diverted to Brazil because there was no immediate prospect of oil production to shuttle to shore until the cleanup was dealt with and normalcy returned to the GoM.

Next came an upheaval in the regulatory regime; in reaction to *Macondo*, a decree by the Obama Administration dismantled the MMS and replaced it with a new regulatory structure led by the "Bromwich Junta." Michael Bromwich was a lawyer who was installed as leader of the new Bureau of Ocean Energy Management, Regulation and Enforcement (BOEMRE). He spoke at a luncheon at the Offshore Technology Conference in May 2011. It was clear to anyone who listened to his hour-long lecture that a new dogma was being installed. MMS was being disbanded, and BOEMRE was taking over. A year or so later BOEMRE split into BOEM and BSEE.

Past regulatory processes that were underway with *Cascade/Chinook* became subject to re-examination and change. With Washington's paranoia in the air, the changing regulatory climate became more difficult for the Petrobras America team, as an operator, to deal with.

Progress came slowly. A chain break on one of the hybrid risers caused more delay until the repair was completed. Along the way, Petrobras America managed the shift from project to steady operation. Their team deserves recognition for their achievement in overcoming adversity!

Recovery and First Oil (2012)

At long last, on February 25, 2012, production operations started in earnest.

Figure 14: Success at Last—First oil at *Cascade/Chinook* Achieved in 2012.
(The FPSO *BW Pioneer* on upper left, at *Cascade/Chinook*)
(Source: BW Offshore)

Operations have continued routinely and successfully ever since. As of early 2016, more than 81 offloadings have taken place with uptimes in the high 90s.

Shell's *Turritella* FPSO

On 24 April 2012, Shell filed its Deep Water Operations Plan (DWOP) for the *Stones* development with the Bureau of Safety and Environmental Enforcement (BSEE). About 200 miles from New Orleans, the development would be located at a record 9,500-foot water depth and would employ an FPSO as a host in Walker Ridge 551. In 2013, Shell announced its firm commitment to the design, construction, and operation of an FPSO under a lease contract with SBM. This became the second FPSO in the U.S. GoM.

In 2013, Shell also committed to a time charter with Overseas Shipholding Group (OSG) for the third shuttle tanker for offloading operations in the U.S. GoM. The *OSG Tampa*, a Handymax products tanker, was delivered from Aker Philadelphia in 2011. In 2014, it was converted in Poland into a shuttle tanker configuration similar to that in Figure 13.

Figure 15: The *Turritella* FPSO on Arrival in the
GoM From Singapore
(Source: Shell)

There are now excellent documents and articles published
(e.g. in Journal of Petroleum Technology, Offshore
Technology Conference, Offshore magazine) on the entire
Stones development, so no attempt is made here to describe
this new FPSO in the GoM, other than with a recent picture
of its arrival in the GoM shown in Figure 15.

Under a ten-year lease commitment with Shell, SBM had
designed and built the *Turritella* FPSO. The management,
design, and construction embraced many advances pioneered
by Shell and Curtis Lohr, the Shell project director for *Stones*
(Figure 16).

Figure 16: Curtis Lohr, Project Director on Stones

Peter Lovie

A Less Serious Side

Almost a year after it was decided that *Stones* would use an FPSO, there came another announcement: its name, *Turritella*. A Shell press release attributed the name to an obscure, little snail.

I had my doubts, though, being familiar with the sense of humor that some Shell employees had. For example, years ago, when people still used Blackberries, I would receive emails from Blake Moore on Shell's *Stones* team. The email signature block read: "Sent from my Blakeberry." There were other humorous cracks, like "getting stoned" on the job. I offered my own contribution: it took many, many meetings in Shell's Houston offices to plan and manage the *Stones* development, commonly meeting around a large U-shaped conference table. In memory of all the gatherings of FPSO pioneers, that conference room needed to be dubbed "Stonehenge" in memory of other historical gatherings.

Pondering the mystery of how the FPSO got its name of *Turritella*, it seemed logical that the FPSO is a ship, thus a "she," hence "Ella." Then it has a turret, so "Turrit." However, I uncovered the bland truth: a committee of project people from SBM and Shell had sifted through 150 possible names and finally arrived upon *Turritella*.

A More Serious Side

The basic achievements of *Turritella* are truly remarkable:

<u>Safest comparable project in Shell and maybe the safest FPSO ever built anywhere: 13.2 million man-hours and not a single incident.</u> Shell describes the combined effort and cohesive team performance of Shell, SBM, and the shipyard with all suppliers. At Keppel's Tuas and Benoi yards, they practiced "the one mindset." It is fascinating to see pictures of Keppel's yard in Singapore because of the elevators and careful scaffolding for efficient work access. It is a far cry from the bamboo scaffolding I'd seen there in the early '70s! What the pictures don't show are the major shifts in yard culture, which emphasized safety and dedication to the construction. These changes paid off in tangible results in the efficiency of time and dollars, as well as intangible results like morale.

<u>Most complete Shell FPSO ever built.</u> Often the rush to get out of the yard translates to construction and commissioning matters still needing attention on location. Their solutions are much more expensive and time-consuming and affect startup schedules.

<u>Combination of design advances in one project.</u> *Turritella* is an FPSO that will work in the deepest waters yet. It is a remarkable combination of engineering achievements. To

simplify and economize production, it employs steel catenary risers in a lazy wave configuration, all while maintaining the ability to disconnect in demanding metocean conditions within the GoM.

Advances in project organization and performance. Shell reduced the number of contractors and suppliers yet preserved competitive performance and good collaboration. There's an unsaid undercurrent of human dignity in play. Over-runs in FPSO projects are legion, but here they have been contained, despite *Turritella*'s groundbreaking features. It is a tribute to how the FPSO contractor (SBM), Shell, and the related suppliers and contractors forged a thought-out working relationship as the project moved ahead.

As in the earlier FPSO ventures in the GoM, there were movers and shakers who quietly and effectively secured the project's success. Curtis Lohr is one of Shell's most experienced FPSO people. From his involvement with the Nigerian FPSO *Bonga* in the 1990s to the *Turritella* development today, Lohr showed leadership in engineering, management safety, and project performance.

The Outlook for More FPSOs in the U.S. GoM

From 2012 to 2013, I served as principal investigator for a U.S. Department of Energy (DOE)-funded project. The project involved a panel of ten GoM operators, who were knowledgeable in the FPSO and offloading business. I asked them whether they saw a need for more FPSOs in the U.S. GoM after Shell's anticipated *Stones* development. The unanimous consensus was that it was unlikely there would be another FPSO contracted in the next ten years. Significantly, this outlook was given before the onset of the severe industry downturn that started in 2H2014.

Multiple field developments that initially looked feasible for FPSOs in GoM ended up selecting another development solution. With the state of the industry in 2016-2018, the prospects look doubly remote because several years of reasonably stable and favorable oil prices are often needed to justify a commitment for a third FPSO in the U.S. GoM.

At the Active Communication International (ACI) FPSO conference in Houston on November 4th-5th of 2015, I was the moderator for one of the two days. In my opening remarks, I put up a slide with FPSO orders worldwide, which showed how they fluctuated widely from 5-25 per year between 2004 and 2014. Two market consultants forecast that there would be 40-50 FPSO contracts worldwide between 2016 and 2020, but those present would have none of it, predicting fewer.

Figure 17 summarizes these trends in annual FPSO orders. After 2008 during the financial crisis, FPSO orders dropped dramatically. Similarly, after the crude price slide started in 2H2014, FPSO orders dropped in 2015. I asked the audience how many thought there'd be four or less FPSO orders worldwide in 2016. There was unanimous agreement on four orders or less in 2016. How about 2017? Still, around a quarter of attendees thought that worldwide orders for FPSOs might still be at the 2016 level, implying most felt things might improve.

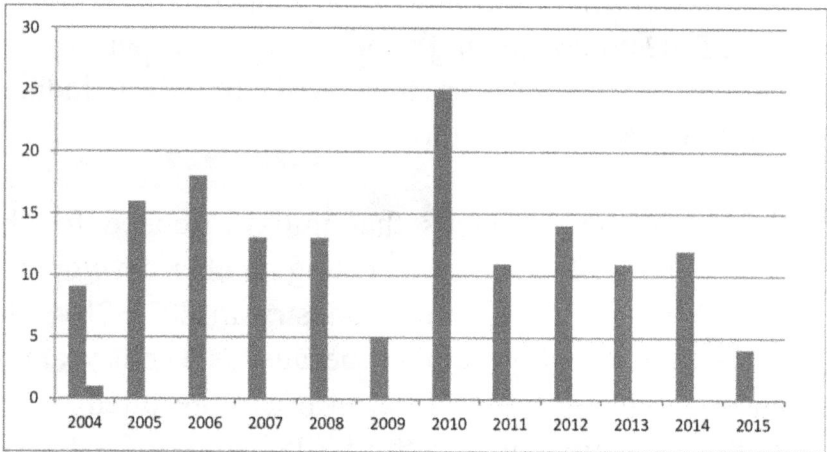

Figure 17: Worldwide FPSO Orders 2004-2015
(IMA, see also Peter Lovie's article in Offshore of May 2015, pp. 108-111)

At another conference, I stated that the FPSO business was beginning to look a lot like the offshore drilling (MODU) business in 1986 and that to keep their great talents and workforces alive, FPSO contractors may have to look to new business lines, such as ocean mining and power barges for other countries, where they could employ their FPSOs and existing technology.

Since November 2015, the crude price has dipped further. It is "lower for longer," implying oil price trends will not make it sufficiently worthwhile for operating oil companies to risk investing in an FPSO. Maybe, as Stephen Guillot noted in November 2015 in *Oil and Gas Investor*, "We are in the downward second half of a twenty-six-year oil price super cycle."

Maybe it is a 1 in 10 chance of another FPSO in U.S. GoM in the next 10 years. Who knows? As they used to say in old movies about gamblers at European casinos in Monaco, coincidentally the headquarters of SBM, "faites vos jeux" (place your bets).

Peter Lovie

Similarities Between the Shuttle Tanker Saga and That of FPSOs

The growth of the shuttle tanker fleet is often overlooked in comparison with that of the FPSO fleet. However, they tend to progress in lockstep. Table C summarizes the history.

Year	Region	Progress
1979	North Sea	First offloading to the *Polytrader* tanker via Bow Loading and taut hawser occurs at Statoil's *Statfjord* development in North Sea.
1982	North Sea	The first DP shuttle tanker, *Wilnora* , starts operation in the North Sea.
1997	Eastern Canada	DP shuttle tankers start operation off Newfoundland with three vessels
2002	Brazil	Petrobras converts two conventional Suezmax (1,000,000 bbl) tankers into DP shuttle tankers for operation in Brazil.
2002	Brazil	First two Teekay DP shuttle tankers (Aframax 700,000 bbl) start operation in Brazil.
2005	Eastern Canada	Two more DP shutle tankers start operation (total: five).
2009	Brazil	Teekay now operates thirteen DP shuttle tankers in Brazil.
2010	GoM	First Jones Act shuttle tanker arrives in GoM but is unable to work because of *Macondo* disaster.
2012	GoM	Shuttle tanker operations finally start in GoM at *Cascade/Chinook*.
2016	GoM	Third Jones Act shuttle tanker enters service at *Stones*.

Table C: Introduction of Shuttle Tankers in Different Parts of the World and now U.S. GoM

The twenty-year saga for the world's shuttle tankers has been one of broad growth. The world fleet grew from the North Sea into Eastern Canada, then into Brazil starting in 2002. Much like the increase in FPSOs, the use of shuttle tankers in Brazil has experienced dramatic growth.

The U.S. GoM is a late and slow entrant, as Table D confirms.

Region	Shuttle tanker fleet	
	1996	2016
North Sea	31	40
Eastern Canada	0	5
Brazil	0	40
GoM	0	3
	31	88

Table D: The Twenty-Year Saga for Shuttle Tankersz

Effect of Allowing Export of U.S. Crude (December 2015)

In December 2015, legislation was signed into law to allow export of crude oil from the U.S. One of the questions that used to be asked was "Can we export the production from an FPSO to St. Croix or somewhere that is outside the U.S.? The answer hitherto had to be no.

Now it would seem there is no reason why an FPSO in the U.S. GoM could not, in theory, offload to a foreign flag export tanker and send the oil outside the country, conceivably with a good financial advantage! However, longer cycle times are a potential drawback, and with shuttle tankers already on charter, there may be little incentive to try foreign tonnage at this point in market history. Still, FPSOs offer Petrobras or Shell new options for their GoM developments.

Peter Lovie

Why Have FPSOs Taken so Long in the U.S. GoM?

There are some fundamental differences between the GoM and the rest of the world which make FPSOs a less favorable developmental solution.

One of these is geography: the U.S. GoM has a flat alluvial plain going out a hundred-plus miles, making it simple and cost-efficient to lay out pipelines to production platforms. This geography stands in stark contrast with, say, the Norwegian Trench, in which the elongated depression of the sea floor helped the prompt development of shuttle tankers in the Norwegian North Sea.

U.S. oil and gas domestic production has been in great demand for U.S. domestic consumption as the country had been an importer for many years. Until the very end of 2015, it was against U.S. law to even export oil from the GoM to other countries. Consequently, there was no incentive to think of storing and sending the oil outside the country, unlike the situation in countries in West Africa, where the need for oil revenues to support their economies is crucial. In these situations, a means of storing oil is essential while it is awaiting tanker export. And what could be easier than storing the crude oil in the same facility that produces it!

Only recently, in a few particularly remote and deep waters

Peter Lovie

in U.S. GoM, has necessity overridden other production and delivery solutions to make FPSOs the ultimate choice.

Closing Thoughts

For its management throughout the tumultuous events in the days before first oil, Petrobras deserves credit. Despite GoM wide events beyond their control such as the hurricanes in 2005 and *Macondo* in 2010, Petrobras persisted toward success and showed leadership, dedication, and teamwork with regulators, the FPSO contractor BW Offshore, and all service providers and installation contractors. *Cascade/Chinook* was truly a precedent-breaking project.

After two earlier runs at using an FPSO in the GoM, Shell pushed the industry limits at *Stones*. Excelling in engineering and making big advances in safety, Shell effectively managed a complex project and achieved great results.

The state of *Stones* in 2016 makes one proud to be an engineer. With today's massive cutbacks and layoffs, *Stones* is a glistening positive.

Shuttle tankers remain a difficult solution for the U.S. GoM. In its requirement for all goods transported between American ports to be carried on U.S.-owned, -operated, and –flagged vessels, the Jones Act shackles the oil industry, despite advances elsewhere in the world. Foreign oil companies benefit from a larger scale of operations. Optimizing and seeing what might be possible with tanker export is illuminating!

Peter Lovie

Even with the effort behind FPSOs in the U.S. GoM, realism seems to have set in; there may be little chance of another *BW Pioneer* or *Turritella* sailing into the GoM within the next decade.

Glossary

Abbreviation	Explanation
ACI	Active Communication International, a London based conference company that held FPSO conferences in Houston in 2012-2015.
AST	American Shuttle Tankers LLC, a Houston based shuttle tanker provider, absorbed into Teekay in 2004.
BLS	Bow Loading System. Shuttle tankers in U.S. GoM are currently using BLS to offload oil from FPSOs into their tanks, for delivery to shore terminals.
BBL	Barrel, a measure of volume usually of oil which equals 42 U.S. gallons.
BLUEWATER	Based in the Netherlands, Bluewater has been one of the original three leading FPSO providers.
BWO	BW Offshore is a Norway based designer, owner and operator of FPSOs, including the first in U.S. GoM.
BOPD	Barrels of oil per day, a commonly applied to daily production of oil.

BOEMRE	MMS was renamed the Bureau of Ocean Energy Management, Regulation and Enforcement (BOEMRE) in mid-June of 2010. On January 19, 2011 BOEMRE was split in two: Bureau of Ocean Energy Management (BOEM) and BSEE.
BSEE	Bureau of Safety & Environmental Enforcement, A U.S. regulatory organization that governs safety onboard the FPSO for handling and processing of oil and gas.
CAPEX	Capital Expenditure.
CONOCO	A U.S. based oil company now known as ConocoPhillips.
DEEPSTAR	An industry group led by Texaco, composed of operating oil companies plus a number of consultants and contractors, all intent on developing needed technologies on a cooperative basis.
DOT 09	Deep Oil Technology conference in 2009.
DP, DP2	Dynamic Positioning in which a vessel holds station with thrusters instead of mooring lines. DP2 is a system with a 2nd system as backup.
DWOP	Deep water operations plan, typically filed with regulators to

	show future use of an FPSO or other type field development solution.
DWT	Deadweight tonnage: total weight of cargo, fuel and supplies in a ship, e.g. a tanker of 100,000 tonnes would be typical of an Aframax tanker.
EIS	Environmental Impact Statement.
EPS	Early production system, the category of floating production system assigned to the first FPSO in the U.S. GoM.
FPSO	Floating production storage offloading vessel, typically ship shaped and frequently converted from a tankers, with oil and gas production facilities and processing equipment on deck.
FSO	Similar to an FPSO but without any production and processing equipment.
HANDYMAX, AFRAMAX, SUEZMAX	Standard sizes of tankers, usually storing about 330,000 bbl, 600,000 and 1,000,000 bbl respectively.
LOOP TERMINAL	Louisiana Offshore Oil Port, an area in U.S. GoM offshore Louisiana where CALM buoys are located to enable offloading of large tankers into pipelines that go

ashore to discharge into storage tanks.

METOCEAN — A category of data for environmental effects affecting loadings on an offshore structures or vessels, e.g. current, wind, and wave.

MEA — Meritorious Engineering Award, awarded annually at OTC by Hart's E&P magazine.

MMS — Minerals Management Service.

MODU — Mobile offshore drilling unit, e.g. a drillship, semisubmersible or jackup.

MOU — Memorandum of Understanding.

NAVION — A Norwegian company created by Statoil in the 1980s then hived off to provide shuttle tanker services, later absorbed into Teekay in 2003.

OTC — Offshore Technology Conference, held annually in Houston since 1969.

OOC — Offshore Operators Committee: a U.S. GoM group serving the interests of operating oil companies in U.S. GoM.

OPA 90 — Offshore Pollution Act of 1990, enacted by the U.S. government after the Valdez tanker spill.

OPEX   Operating Expenditure.

OTRC   Offshore Technology Research
                     Center: located in College Station,
                     Texas at Texas A&M University and
                     active in the EIS.

SPOT MARKET   The rate for buying or selling
                     hydrocarbons— or for hiring a
                     vessel— is set at the day of contract
                     signing at whatever level the
                     market of many competitive offers
                     indicates rather than negotiated in
                     advance.

SBM   SBM Offshore N.V. in the
                     Netherlands, with a large Houston
                     office, is an industry leader: first
                     designer and builder of an FPSO
                     in 1977 offshore Spain and designer
                     and builder of *Turritella* FPSO at
                     the *Stones* development in U.S.
                     GoM.

SHELL   The supermajor oil company
                     with a large Houston office
                     that is operator on the *Stones*
                     development and recently became
                     owner - of the *Turritella* FPSO, the
                     second in U.S. GoM.

SPE   Society of Petroleum Engineers.

S-S-S   *Separate storage shuttling*, a
                     concept devised by American
                     Shuttle Tanker whereby a
                     dynamically positioned tankers
                     stores production from an adjacent

	non FPSO field development, for offloading to shuttle tankers.
STATOIL	A leading Norway based oil company that pioneered shuttle tanker offloading in the North Sea in 1979, known as Equinor since early 2018.
TEEKAY	Teekay Corporation based in Vancouver is a leading provider of shuttle tanker and FPSO services.
TEXACO	A major oil company, leader of the EIS effort and of Deepstar, now integrated into Chevron.
ULCC	Ultra large crude carrier, the largest category of tanker, typically with storage capacities of 2.4 to 4 million barrels of oil.
USCG	U.S. Coast Guard. An arm of the U.S. military and the regulatory authority vetting safe marine operations, e.g. offloading from an FPSO to a tanker and movements of FPSOs and tankers.
VLCC	Very large crude carrier, typically 2,000,000 bbl storage capacity.

Appendix A: Relevant Publications

These documents are cited in chronological order and give some industry context. They are freely downloadable at www.FPSOsinGoM.com

1. Lovie, P.M.: "The Lower Tertiary Trend and the Oil Export Economic Prize", Deep Oil Technology (DOT), conference, paper 138, New Orleans, 3 February 2009, 28 pages.

    Sets forth the results of studies during 2007-2009 by Devon Energy on the cost effectiveness of pipeline and shuttle tanker export in deep more remote waters of U.S. GoM. Compares use of FSOs or FPSOs in these field developments.

2. Lovie, P.M.: "An Independent Look at the Round Hull Concept", 7th FPSO Vessel Conference, Active Communications International (ACI), Houston, 4-5 November 2015, 33 slides.

    A discussion of round and shipshape FPSOs. It offers both serious and entertaining comparisons. Dr. Irving Finkel's discovery that Noah's Ark circa 2400 BC was round and Peter Lovie discovery that it was the same diameter as the first round FPSO in 2006 combine to lead to interesting conclusions. Perhaps today's patents on round-hull MODUs and FPSOs face "prior art" of truly biblical proportions!

3. Lovie, P.M.: "2016 and Two FPSOs in US GoM: The

Twenty Year Saga", presentation at Rice Global E&C Forum, Roundtable meeting 10 June 2016, 47 slides.

This is a PowerPoint presentation on some of the content in this book, made to an audience of construction contracting professionals at Rice University.

4. Dittrick, P.: "Shell Stones field advances subsea production technology", Oil & Gas Journal, 1 May 2017 online, 6 pages.

A leading petroleum industry publication cites Peter Lovie's views on FPSOs in the U.S. GoM.

Appendix B: The *Upstream* Interview of Peter Lovie

New FPSOs in new parts of the world attract industry attention, no more so than in *Upstream*, a weekly newspaper that is read worldwide by movers and shakers in the offshore and upstream world.

Every week, *Upstream* publishes an interview with someone they see as a leader in the industry, sometimes a U.S. Senator, an OPEC official, an industry CEO … or a pioneer like Peter Lovie. What follows is the *Upstream* interview of Peter Lovie on July 22, 2016, included here instead of a biography of the author.

Trailblazer: Peter Lovie took a six-foot model of the Glas Dowr floater up and down the towns of the US Gulf coast as he campaigned to bring FPSOs to the region
Photo: KATHRINE SCHMIDT

FPSOs — a labour of Lovie

Meet the man who brought **FPSOs to the US offshore** — by **dragging a six-foot model of one** up and down the coast

Peter Lovie

KATHRINE SCHMIDT
Houston

PETER Lovie remembers well his late 1990s road trips to the towns of the US Gulf coast with an unusual companion — a six-foot model of the Glas Dowr floating production, storage and offloading unit.

The veteran engineer took the unit on the back of a truck to places from Corpus Christi to Lafayette as a key part of an environmental impact study that paved the way to bring FPSOs to the US offshore.

Bringing the technology to the region has been a long and winding road, a tale that Lovie recently wrote in a history entitled "Two FPSOs in the Gulf of Mexico: A 20-year saga".

The industry's effort over time culminated in the start-up of the Petrobras-operated Cascade-Chinook development in 2012 and Shell's Stones project, which is due on line soon.

"It's the kind of project that makes one proud to be an engineer," he says. "It's something inspiring to the industry."

A native of Fife, Scotland, Lovie took a civil engineering degree at the University of Glasgow and a master's at the University of Virginia as a Fulbright scholar.

Cameron recruited him to work in Houston, where he moved to a position at the predecessor to Transocean and ran engineering outfits focused on jack-up rig design and later subsea processing.

In 1995 Lovie joined FPSO provider Bluewater and was tasked with business development in the US Gulf.

Pioneering vessels By then, about 50 units were active around the world as pioneering vessels such as the Castellon off Spain in 1977 gave way to harsh-environment units in the North Sea.

In 1996, Texaco considered one of the first US Gulf FPSO projects for its Fuji prospect, then a remote deep-water well in about 1700 feet of water.

However, regulators first wanted an environmental impact study on the field development model amid concern following the 1989 Exxon Valdez spill in Alaska.

Industry collaborated through the DeepStar technology alliance to complete the document, which took $3 million and five years.

Following his Bluewater post, Lovie began working on FPSO projects from another perspective, in 2002 joining American Shuttle Tankers, later acquired by Teekay.

In his various roles, Lovie had a front-row seat for the ongoing field development discussions on projects that considered FPSOs.

From 1998 to 1999 Shell weighed up the options for its Na Kika development, but ultimately opted for a semi-submersible design.

The operator struggled to crack the problem of risers, later mastered in Brazil's BM-S-10 block with the lazy wave steel catenary design.

In 2000, BP considered a shuttle-tanker option for what later became the Mardi Gras pipeline system for its flagship fields in Mississippi Canyon and Green Canyon, such as Thunder Horse, Mad Dog and Atlantis.

Shell in 2004 also contemplated an FPSO at its Great White discovery, which later became the Perdido spar.

However, it finally selected the FPSO model for its Stones development, which is set to be produced by the SBM Offshore Turritella unit, poised to become the world's deepest production facility in 9500 feet of water.

The shuttle tanker question was central as operators contem-

> To celebrate becoming a US citizen, he ordered a pair of customised Western boots with the US seal.

plated the FPSO model, and was complicated in the US by the Jones Act, which requires goods transported between two US ports to be carried by US built, flagged and staffed vessels.

US tankers were more expensive to build, and there was no guarantee that the life of the field would correspond with the life of the vessel, factors that weighed on development economics.

Disconnectable turret The 2005 hurricanes Katrina and Rita forced an industry rethink of design properties on offshore facilities and, in the case of FPSOs, spurred the idea of a disconnectable turret to avert collisions and spills during storms.

In 2006, Lovie joined US independent Devon Energy to work on its Lower Tertiary deep-water developments, where it was partnered with Brazilian state oil company Petrobras on Cascade-Chinook in the Walker Ridge area.

The FPSO concept won the day amid uncertainty about how the Lower Tertiary find might produce over time.

The company also handled tanker economics by learning to live with a higher dayrate and constructing vessels with some, but not all, of the manoeuvrability of top-flight dynamic positioning units. The BW Pioneer FPSO began producing at the fields in 2012 after the Macondo disaster, at the time the world's deepest such development in 8200 feet of water.

Lovie left Devon in 2009 and has since worked as an independent consultant.

He has also led and moderated dozens of offshore conferences over the years, and led a US Department of Energy research paper on deep-water offloading.

Despite the progress, Lovie sees an uncertain outlook for future FPSOs. Redeployment of existing units can be harder than it looks, and operators tend to underestimate the costs of adaptation.

"Each... has been built for a particular oilfield, and each oilfield is different," he says.

Lovie lives in Houston with his wife and has a grown son. He likes to delve back to his roots by travelling back to Scotland or savouring a good single malt such as Laphroaig or Talisker.

However, he has also fully embraced his American home and became a US citizen in 2008.

To celebrate, he ordered a pair of customised Western boots with the US seal that he proudly wears to business meetings.

He recalls the citizenship ceremony alongside 2500 others, all dressed in their best, representing about 100 nationalities and speaking myriad languages.

"A truly unforgettable experience, moving," he says. [U]

Peter Lovie